ACKNOWLEDGMENTS

Thanks to Ron Hemeon,
who provided the inspiration for
this story in the form of a
strikingly handsome blue lobster
named "George."

Thanks to Doris Conway, Kristen Petrie, David Petrie,
Samantha Romage, Sally MacLaughlin, Cheryl Petrie,
Stephanie McDermott, and Mike Petrie for their advice
and being my sounding board in support of this project.

This book is dedicated to Steve and Thelma Connolly

To Mike, whose belief remains steadfast:
Here's to soaring with the eagles!

Copyright: September 2015 by Janice S. C. Petrie
Seatales Publishing Company

All rights reserved. No part of this book may be
used without the written permission of Seatales
Publishing Company or the author. Printed in
the United States of America.

ISBN 10: 0-9705510-3-7

ISBN 13: 978-0-9705510-3-0

Something's Tugging on My Claw!

Written and Illustrated by
Janice S. C. Petrie

Something's tugging on my claw.
It could be friend or foe.
It pulls, and pulls, and pulls some more,
and it just won't let go!

I think I'd better drop my claw.
It's what we lobsters do.
When danger's near, it's very clear;
one claw is better than two.

I'll hide away, both night...

and day,
until the danger's gone.

But when I'm really hungry,
I'm afraid I must move on.

He's looking at a lobster trap, and going in. Hurrah!

The swordfish sees an octopus,
and takes off real fast towards it,
into a cave; he's very brave.
But I'm not brave, not one bit!

I'll get away, no need to stay.
I'll flap my tail so swiftly;

to back away, no time to play,
I must go backwards quickly!

Oh, no! I've found a horseshoe crab.
Did she tug on my claw?

Of course not, little lobster.
It wasn't her I saw.

I saw a man tug on your claw,
with its gleaming cobalt color.
You don't blend well, with your cyan shell.
Your color should be duller.

Stay far away from the shoreline.
That's what I'd advise.

Horseshoe crabs hide in the sand.

Some fish hide in the sand, too. Fish blend into the sand to rest or hunt.

Shush, I think I had a nibble!

Or dig yourself into the sand.
I've always thought that wise.

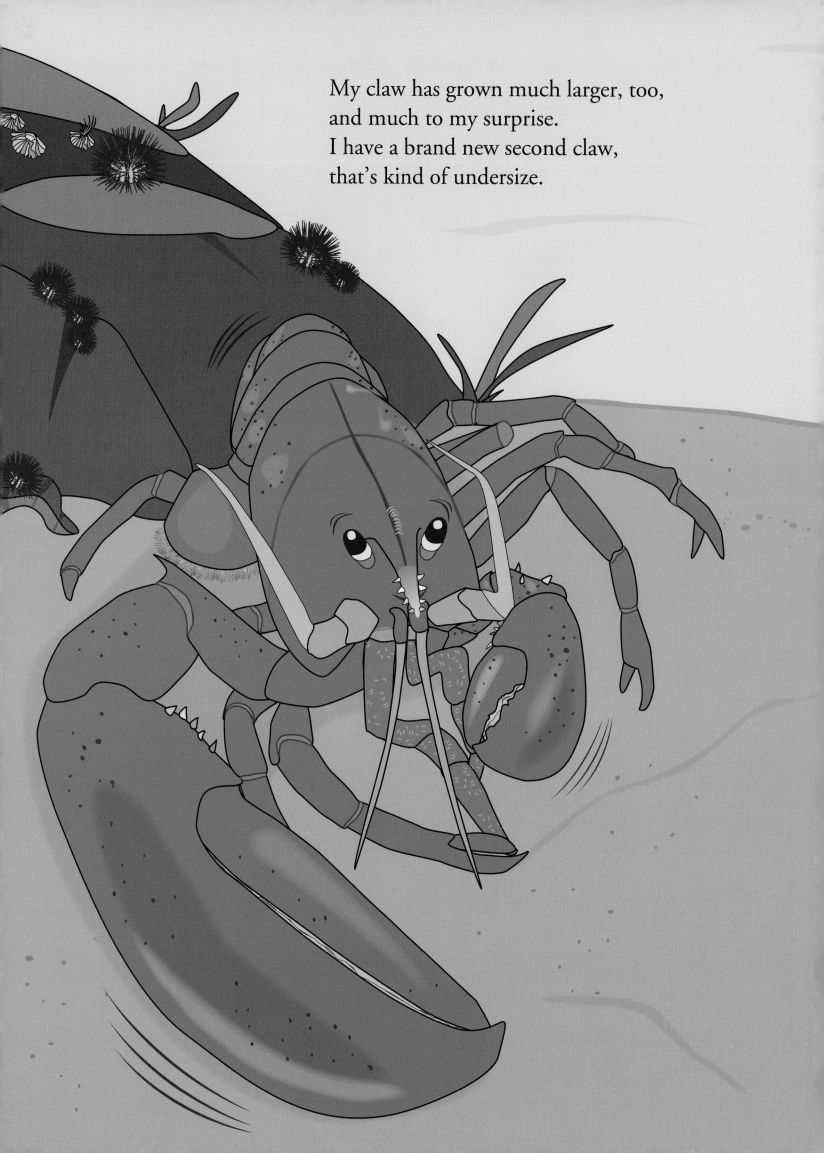

My claw has grown much larger, too,
and much to my surprise.
I have a brand new second claw,
that's kind of undersize.

I finally have two claws again,
a pincher and a crusher.
And now I'm very wise to keep
my blue claws under cover.

Lobster Facts:

Not All Claws Are the Same

Take a close look at an American lobster's two claws and you'll see that each claw is designed with a unique purpose in mind. The pincher claw is for tearing her food, while the crusher claw is designed to hold and break the lobster's food. Some lobsters have their crusher claw on their left side, and their pincher claw on the right. These lobsters are considered left side dominant, or left-handed. Other lobsters have their crusher claw on their right side, and their pincher claw on their left. These lobsters are thought of as being right side dominant, or right-handed. A lobster with only one claw, like the blue lobster in this story, is called a cull. An American lobster that's lost both her claws is called a pistol. When a lobster's claw is held tightly and the lobster feels threatened, she'll release her claw from her body, to protect herself from being captured.

Does a Lobster Trap Really Have a Kitchen & a Parlor?

Any lobster fisherman can tell you that most lobster traps of today have a funnel that serves as the entrance to the trap, a kitchen where the bag full of food or bait can be found, and a parlor for the lobsters to stay in until the trap is hauled up into the boat. The number of kitchens and parlors can vary from trap to trap, but all traps must have an escape vent to allow undersized lobsters to get away. A funnel is not only used as an entrance to the trap, but additional funnels are used as an entrance to each room inside the trap.

According to Maine's Department of Marine Resources, in order for a lobster to be legally harvested, he must be no shorter than 3 1/4 inches and no longer than 5 inches when measured from the rear of the eye socket, straight down the center of his body, to the rear end of his body's shell, not including his abdomen or tail. It's common for lobster fishermen to find Jonah crabs in their lobster traps, because lobsters and Jonah crabs enjoy living in the same habitat.

How Can You Tell a Boy (Male) Lobster from a Girl (Female)?

Male lobsters have a narrower tail than a female lobster of the same age. The female lobster has a wider tail because when she has eggs, she keeps them on the underside of her tail where her swimmerets are located. Males don't have eggs, so they don't need all that extra room in their tail. Also, if you turn a lobster on its back and look at the first set of swimmerets near where its body and tail meet, the female's first set of swimmerets will look and feel soft, much like her other swimmerets. The male's, which are called gonopods, will be as hard as his shell, and larger than the female's first swimmerets.

Male (Boy)

Female (Girl)

Why Are Some Lobsters Blue?

Although there are many theories as to why some lobsters are blue, the most widespread belief is that blue lobsters have a genetic defect that cause their bodies to produce more of a certain protein than the ordinary greenish, blackish, brown lobsters. Although considered very rare, it's thought that blue lobsters are found in every one in two million caught. Believe it or not, it's much more rare to find a lobster that's alive and red, yellow, or calico. And if a rare albino lobster is caught, whose shell is considered colorless, appearing almost like crystal, that's a one out of 100 million find.

A Word About Molting

When lobsters grow, they molt and shed their shell, just like a snake sheds its skin to grow. Once the lobster crawls out of its old shell, its very vulnerable to predators. For this reason, he'll hide under a rock or seaweed, and eat his old shell while he waits for his new shell to become hard. The old shell contains nutrients that when eaten by the lobster, helps his new shell to harden quicker. Each time the lobster molts, his new claw will grow larger.

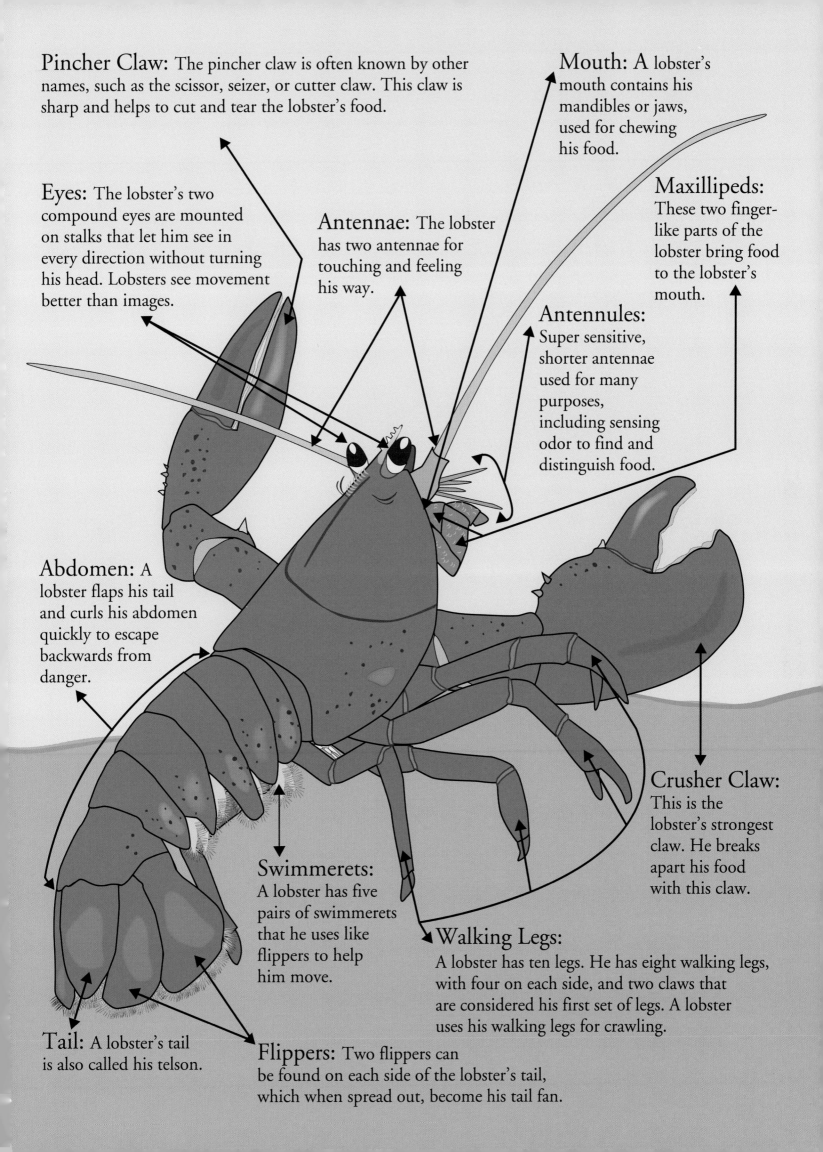